Fields of Faith

Scriptures and Prayers for Farmers

Sallie Dawkins

Fields of Faith: Scriptures and Prayers for Farmers

Library of Congress Control Number: 2025923363

eBook ISBN 978-1-955861-64-9; Print Paperback ISBN 978-1-955861-62-5

Published by
Open Ark Books
and Firebrand United, LLC
216 Skywatch Dr. PMB 206
Danville, Kentucky
40422 USA

Author Website:
www.SallieDawkins.com

Original text and photographs by Sallie Dawkins. Image conversion to storybook effect using ChatGPT5.

For Kyle

Thank you for helping me to see the wonder of the world through your eyes.

"This is the day
the Lord has made.
Let us rejoice
and be happy today!"

Psalm 118:24

Thank You, God,
for family farms,
where hard work
cultivates
and love warms.
May our hands be strong
and hearts be pure,
in Your presence,
we feel secure.

"The earth is
Yahweh's,
with its fullness;
the world,
and those who
dwell in it."

Psalm 24:1 WEB

In springtime's warmth,
we sow our seeds,
with joy and hope,
our hearts take heed.
God bless the rain,
the sun so bright,
and keep our crops
from drought or blight.

"You make the
grass grow to
feed the animals.
You provide plants
for the crops we
grow- the plants
that give us food
from the earth."

Psalm 104:14

Thank You, God,
for all You give,
for every blessing
that helps us live.
As we sow the seeds
of kindness today,
guide our hearts,
show us the way.

"Let's not be weary
in doing good,
for we will reap
in due season
if we don't give up."

Galatians 6:9 WEB

Lord, grant us wisdom
to know our part,
in caring for creation
with an open heart.
May every choice
reflect Your decree,
in harmony with nature,
may we always be.

"The Lord
will bless you
and fill your barns.
He will bless
everything you do.
The Lord your God
will bless you
in the land that
He is giving you."

Deuteronomy 28:8

With spring's sweet song
and summer's glow,
may our blessings
blossom, and
our gardens grow.

"The one who plants
and the one who
waters have the same
purpose. And each
one will be rewarded
for his own work."

1 Corinthians 3:8

Dear God,
thank You for
the fields
and the fruits
they give.
Thank You for
sunshine and
showers of rain.
Cause kindness
to grow in our hearts
as we seek You
and share Your love
with others.
Amen.

"In everything
give thanks,
for this is the
will of God
in Christ Jesus
toward you."

1 Thessalonians 5:18 WEB

Thank You, Lord,
for Your endless love,
You fill our lives
and overflow
Your blessings
from above.

"Shout to the Lord,
all the earth;
break out in
praise and
sing for joy!"

Psalm 98:4 NLT

For the morning sun,
and the seeds we sow,
we thank You, Lord,
for fields that grow.
In every bud and
bright green leaf,
Your goodness blooms
beyond belief.

"In all the work you
are given, do the
best you can.
Work as though
you are working
for the Lord, not
any earthly master."

Colossians 3:23

May we always
appreciate
the beauty of nature,
and the food that
nourishes us,
dear Creator.

"May mercy, peace,
and love be
multiplied to you."

Jude 1:2

In every season,
through joy
and strife,
may we feel
Your love, Lord,
in the dance
of life.

"If you plant goodness, you will harvest faithful love. Plow your ground, and you will harvest with the Lord. He will come, and He will make goodness fall on you like rain."

Hosea 10:12

When autumn's chill
brings harvest time,
we gather fruits,
a gift sublime.
With thankful hearts,
we share our yield,
God's bounty blessed
upon this field.

"The Lord says, 'You must listen carefully to the commands I give you today: You must love the LORD your God, and serve Him with all your heart and all your soul. If you do that, I will send rain for your land at the right time.'"

Deuteronomy 11:13 ERV

As sunflowers stretch
to greet the day,
may gratitude guide us
on our way.
With open hearts
and charitable hands,
we'll share our blessings
across the lands.

"Be strong, all you people of the land," says Yahweh, "and work, for I am with you," says Yahweh of Armies.

Haggai 2:4 WEB

With each little act,
may the world see,
Your love and grace,
with us, shall be.

"Look at the
farmers. They
have to be patient.
They have to wait
for their valuable
crop to grow and
produce a harvest.
They wait patiently
for the first rain
and the last rain."

James 5:7

Father,
we trust in
Your promise
of good harvests.
Thank You
for the land
You've given us.
Help us
to work hard
and celebrate
the fruits
of our labor!
Amen.

"So begin planting
early in the morning,
and don't stop
working until
evening. You don't
know what might
make you rich.
Maybe everything
you do will be
successful."

Ecclesiastes 11:6

Father God,
thank You,
for guiding us
in our work.
Help us be patient
and to trust
in Your seasonal
provisions.

"As long as the earth continues, there will always be a time for planting and a time for harvest. There will always be cold and hot, summer and winter, day and night on earth."

Genesis 8:22

For quiet skies and
resting earth,
we thank You, Lord,
for each day's worth.
Though fields lie still
and branches bare,
Your faithful love
is everywhere.

"The LORD of
Heaven's Armies
is a wonderful
Teacher,
and He gives the
farmer great
wisdom."

Isaiah 28:29 NLT

May our hearts
be filled with thankfulness
for the simple things,
like the rain
that helps plants grow,
and the food
that brings us joy.

Harvest
Blessings

"People who work
for peace in
a peaceful way
get the blessings
that come from
right living."

James 3:18

Teach us, Lord,
to always listen
to Your commands
and share our hearts
with others,
so that they too
may feel
Your goodness!

"God is the one who gives seed to those who plant, and He gives bread for food. And God will give you spiritual seed and make that seed grow. He will produce a great harvest from your goodness."

2 Corinthians 9:10

Fill our tables
with Your light,
Help us trust You,
day and night.
As we gather
the bounty
of the land,
We thank You, Lord,
for Your
generous hand.

"Let the favor of
the Lord our God
be on us. Establish
the work of our
hands for us. Yes,
establish the work
of our hands.."

Psalm 90:17 WEB

Let us move
forward,
enriched by faith,
to grow
as caretakers
of the land
and good stewards
of God's blessings.

"Commit your work to Yahweh, and your plans will be established."

Proverbs 16:3 WEB

Heavenly Father,
Thank You
for the land,
the sunshine,
and the rain.

Bless the farmers
who plant and harvest,
and bless our families
as we share food
together.

Help us grow
in kindness,
planting seeds
of love
wherever we go.
Amen.

EXTENDING THE STORY

DRAW IT: Draw a farm scene with fields, animals, or a barn. Add sunshine or rainclouds to remind you of God's blessings.

COUNTING BLESSINGS: How many fruits or vegetables can you name? Which ones are your favorites?

FAMILY TALK: Ask your parents or grandparents if they ever grew food. What did they learn? What season is it right now? What do farmers do in this season?

PRAYER TIME: Write or say a short prayer thanking God for farmers and the food they help provide.

NEXT STEPS

The illustrations for this story were created from photos taken from the author's time on a real farm.

If you enjoyed this story, please consider leaving an honest review on Amazon or Goodreads. Your words have the power to inspire other readers, and spread the important message of love and support this book carries.

To connect with the author, visit www.SallieDawkins.com.

Made in United States
North Haven, CT
18 October 2026

20502738R00053